GEMS, CRYSTALS, AND PRECIOUS ROCKS

Steven M. Hoffman

PowerKiDS press.

New York

Published in 2011 by The Rosen Publishing Group, Inc.
29 East 21st Street, New York, NY 10010

First Edition

Editor: Amelie von Zumbusch
Book Design: Kate Laczynski
Layout Design: Ashley Burrell

Photo Credits: Cover, pp. 4 (bottom right), 12 (right) Ken Lucas/Getty Images; p. 4 (top left) DEA Picture Library/Getty Images; p. 4 (top right) Hemera/Thinkstock; p. 4 (bottom left) © www.iStockphoto.com/Stefan Witas; p. 6 Jeffrey Hamilton/Digital Vision/Thinkstock; p. 8 © www.iStockphoto.com/Maria Barski; p. 10 Cesar Lucas Abreu/Getty Images; pp. 12 (left), 16 Shutterstock.com; p. 14 18th Dynasty Egyptian/Getty Images; p. 18 Lihee Avidan/Getty Images; p. 20 © www.iStockphoto.com/Ugurhan Betin.

Library of Congress Cataloging-in-Publication Data

Hoffman, Steven M. (Steven Michael), 1960-
 Gems, crystals, and precious rocks / by Steven M. Hoffman. — 1st ed.
    p. cm. — (Rock it!)
 Includes index.
 ISBN 978-1-4488-2561-5 (library binding) — ISBN 978-1-4488-2708-4 (pbk.) — ISBN 978-1-4488-2709-1 (6-pack)
 1. Precious stones—Juvenile literature. 2. Crystals—Juvenile literature. 3. Gems—Juvenile literature. I. Title.
 QE392.2.H64 2011
 553.8—dc22
                                        2010031831

Manufactured in the United States of America

CPSIA Compliance Information: Batch #WW11PK: For Further Information contact Rosen Publishing, New York, New York at 1-800-237-9932

CONTENTS

| Aquamarine | Diamond |
| Amber | Lapis Lazuli |

Aquamarines and diamonds are two gems that are minerals. The gem amber is made of fossilized tree sap, while the gem lapis lazuli is a rock.

Beautiful Stones

Have you ever seen a gem? Gems are beautiful, hard-to-find stones that can be used in **jewelry**. When they have been cut, many gems flash brightly in the light. Some have beautiful colors. Gems are often used in rings, watches, and necklaces. Some of the largest have been included in the crowns of kings and queens.

Most gems are **minerals**. Minerals are solids. They are also natural crystals. A crystal has an ordered pattern of **atoms**. Some crystals have flat crystal faces.

There are a few gems that are not mineral crystals. Many of these are rocks. Rocks can be made up one or more minerals.

These gems are diamonds. Diamonds are known for their sparkle. They are also the hardest mineral and the hardest thing found in nature.

Many Types of Crystals

Crystals are common in nature. Sugar and salt are crystals. Crystals of the mineral quartz are used to make clocks. Diamonds, emeralds, sapphires, and rubies are just some of the gems that are crystals.

Crystals are grouped into six groups based on the patterns that their atoms form. The groups are known as crystal systems. Each crystal system has several related crystal shapes in it.

For example, diamonds belong to a crystal system called cubic crystals. Diamond crystals may have the shape of a **cube**. They may also have other shapes. In fact, the most common shape for a diamond is a crystal that has eight triangular faces.

The salt crystals on this rock formed when salt came out of the waters of the Dead Sea. The Dead Sea is between Jordan, Israel, and the West Bank. It is very salty.

Making Crystals

Crystals form in several ways. The first is from **magma**. Magma is melted rock that forms inside Earth. When magma cools, it forms crystals. Small crystals form when magma cools quickly. Large crystals form when it cools slowly.

Crystals can also form from matter dissolved in water. **Salts** form in this way. Many gems, such as amethyst and fluorite form this way, too. Crystals may form when salty water dries up. They also form as hot water inside Earth cools.

The third place that crystals often form is in hot rock. When rock is deep under ground or near magma, heat and **pressure** can form new crystals in the rock.

Emeralds always have a rich, green color. These emeralds are from Colombia. Many fine emeralds come from Colombia.

Gem Crystals

Crystals of some minerals are used as gems. To be used as a gem, crystals must be very beautiful. The crystals also must have good color and clearness. Most of the crystals used as gems are rare.

A diamond is a good example of a gem crystal. Diamonds are often clear. They can also have beautiful colors, such as blue, pink, or yellow. When they have been cut and **polished**, diamond crystals sparkle in the light. Diamonds are also very rare. They are found only in rocks from certain kinds of **volcanoes**.

Diamonds form deep inside Earth. However, these gems have been carried to Earth's surface in some places by rising magma.

The blue gems in the rock on the left are sapphires. The red gem on the right is a ruby. While rubies have chromium in them, blue sapphires have titanium.

Different Kinds of Gems

Gem crystals of the same kind of mineral sometimes have different colors. When this happens, the crystals may be known by different names. The gems ruby and sapphire are both crystals of the mineral corundum. The difference is that rubies are red and sapphires are generally blue. Small amounts of different atoms cause these different colors. For example, rubies have some atoms of a metal called chromium in them.

Crystals of a mineral called beryl also occur in many colors. Dark green gems of this mineral are called emerald. Bluish green gems are aquamarine. Yellow ones are golden beryl.

Some rubies and sapphires show stars of light when looked at from certain directions. These are called star rubies and star sapphires.

The blue stones in this necklace are lapis lazuli. The necklace was made in Egypt over 3,000 years ago.

Gems from Rocks

Not all gems are mineral crystals. Some, such as obsidian, are rocks. Obsidian is volcanic glass. It forms when lava cools too quickly to form crystals. Obsidian is often black but may be other colors. Some obsidian has a snowflake pattern in it. This happens when some small crystals form in the glass. These gems are called snowflake obsidian.

Lapis lazuli is a beautiful blue gem. It is cut from a rock that includes crystals of several different minerals. Some lapis lazuli has small crystals of the mineral pyrite in it. Pyrite is also known as fool's gold. These pyrite crystals shine brightly in the blue stone.

Pearls form when something gets inside the shell of certain animals, such as this oyster. That thing gets covered with layers of the same matter that coats the inside of the animal's shell.

Organic Gems

Organic gems are gems that come from animals or plants. Pearls are the best known of these. Pearls form inside the shells of certain clams, such as the *akoya* oyster. Pearls are often light in color but may be darker, too. Black pearls are highly valued. Each pearl has many layers that separate light into a rainbow of colors.

Other organic gems include amber and jet. Amber is hardened tree sap. It is usually yellow, orange, brown, or red but can be other colors, including green. Amber is often polished and used in necklaces or rings. Jet is a gem that comes from a type of coal. It is most often black and very shiny.

These men are looking for sapphires in a stream in Madagascar. Many of the sapphires mined today come from Madagascar.

Hunting and Mining Gems

People hunt for gems in many ways. They often start by flying over land in airplanes. They bring along special machines to study the land. The machines measure rock **properties**, such as small differences in **magnetism**. Then, miners move to the ground. They study the rocks and soil. If they find a few gems, they look for others nearby.

After people find spots that are rich in gems, they dig mines. The mines might be deep tunnels. They could also be large open pits. Many gems are mined from streams. Miners sometimes use ribbed boxes called **sluices** to trap the gems. They also use open pans like those used to mine gold.

This jeweler is setting a gem. This means the jeweler is placing the gem in a metal frame that will hold it in place.

Finishing Rough

To be used in jewelry, gems must be cut and polished. First, the uncut stones, also called rough, are cleaned of any dirt or rock. If a gem is large, a jeweler might then cut it into smaller pieces that will make good finished stones.

The gem is then cut using a special grinding machine. The jeweler may cut curved surfaces or flat surfaces, called **facets**, into the gem. The last step is to polish the gem. This makes the surfaces very smooth. It is done with a fine polishing powder. Finally, the gem is ready for a ring or necklace.

Gems cut to have curved top surfaces and, often, flat bottoms are called cabochons. Gems cut with many facets are faceted stones.

Learning About Gems

You can see and learn about gems in many places. Gem shows are yearly events in many towns and cities. People who are interested in gems gather to look at stones and share stories.

There might also be a rock store near you that sells gem rough or even finished stones. The people at these stores often know a lot about gems.

Many people know someone who has a diamond ring or pearl earrings. To see the largest gems, though, you will have to visit a museum. The Smithsonian Museum in Washington, D.C., is one of the best. You will be wowed by the large, blue Hope Diamond and the other gems held there.

GLOSSARY

atoms (A-temz) The smallest parts of elements.

cube (KYOOB) A shape with six square sides.

facets (FA-sets) Smooth sides cut into a gem.

jewelry (JOO-ul-ree) Objects worn for decoration that are made of special metals, such as gold and silver, and prized stones.

magma (MAG-muh) Hot, melted rock inside Earth.

magnetism (MAG-nuh-tih-zum) The force that pulls certain objects toward a magnet.

minerals (MIN-rulz) Kinds of natural matter that are not animals, plants, or other living things.

polished (PAH-lisht) Rubbed something until it shines.

pressure (PREH-shur) A force that pushes on something.

properties (PRAH-pur-teez) Features that belong to something.

salts (SAWLTS) Solids formed from particles that have opposite electrical charges.

sluices (SLOOS-ez) Human-made paths for water.

volcanoes (vol-KAY-nohz) Openings that sometimes shoot up hot, melted rock called lava.

INDEX

WEB SITES

Due to the changing nature of Internet links, PowerKids Press has developed an online list of Web sites related to the subject of this book. This site is updated regularly. Please use this link to access the list:
www.powerkidslinks.com/rockit/gems/